P9-DNN-098

Kunst und Architektur im Gespräch | Art and Architecture in Discussion

edited by
Cristina Bechtler
in collaboration with Kunsthaus Bregenz

Kunsthaus Bregenz

Pictures of Architecture
Architecture of Pictures

Architecture of Pictures

A Conversation between
Jacques Herzog and Jeff Wall,

moderated by
Philip Ursprung

CONTENTS

PREFACE

Contemporary architecture has increasingly acquired cult status as a much-discussed topic in all the media. Urban planners in particular welcome spectacular contemporary buildings as a means of drawing attention to their cities and enhancing the cultural appeal of entire regions. Well-placed, first-class luxury buildings not only change the image of entire neighbourhoods; they also exert a positive influence on the local economy. The importance of excellence in architecture is of growing significance worldwide. This applies to contemporary art as well, although the latter does not enjoy the widespread appreciation of its more popular companion. Contemporary art is more demanding and its meanings more intricate; it eludes easy legibility and engages in complex reflections on social reality.

Architects often take inspiration from contemporary art, not only from its haptic, material presence and imaginative treatment of materials, but also from its analytic examination of society. Art and architecture have once again entered into a mutually fruitful dialogue. Innovative architecture proposes solutions that incorporate artistic strategies; conversely, the subject matter of art may often be articulated in relation to architectural givens.

The series *Art and Architecture in Discussion* has been launched to provide a forum for debate on the conflicts and tensions that often accompany and enrich the interaction between the two fields. The present publication provides a platform of discussion for two major representatives of international art and architecture, who know each other's work well, although they have not previously collaborated on concrete projects.

In his oeuvre, Jeff Wall makes reference to art history as well as examining social and political issues and the role of photography within the visual media. Since the 1970s, he has worked with photography in the form of large-format transparencies presented in light boxes. Photography still lays claim to mirroring a specific reality. Jeff Wall's pictures tie in with that claim inasmuch as they sometimes resemble snapshots while actually being the result of elaborate planning in which no detail is left to chance. The intense luminosity of the backlit photographs invests them with a pictorial reality of their own.

Jacques Herzog, who started out in the fine arts himself, is highly interested in Jeff Wall's artistic strategy. Wall has, in turn, repeatedly addressed the motif of architecture, as in his photograph of Herzog & de Meuron's Dominus Winery in California, which was not commissioned by the architects, as one might assume, but by the Canadian Centre for Architecture in Montreal.

Since the 1980s, Herzog & de Meuron have designed and executed a number of projects specifically for the arts (Tate Modern and Laban Dance Centre, London, Walker Art Center, Minneapolis, Schaulager, Basel and Kunsthaus Aarau). Their buildings often emerge in close collaboration with artists like Rémy Zaugg, Thomas Ruff and Adrian Schiess, with whom they have worked several times. The largest project, currently under construction, the stadium in Beijing for the 2008 Olympics, has already been dubbed the "Swallows Nest". The architects are working together with the Chinese artist Ai Wei Wei, who has described their approach as follows: "They naturally relate their work to intelligence, art and philosophy, and they associate themselves with the dream of the freedom of the human condition in a ruthless but restricted way."

Jeff Wall, Cristina Bechtler, Jacques Herzog, Philip Ursprung,
November 4th 2003, office of Herzog & de Meuron, Basel

This conversation between Jeff Wall, Jacques Herzog, myself and Philip Ursprung as moderator took place on the 4th of November 2003, from 4 – 8 p.m., at the offices of Herzog & de Meuron in Basel. I am very happy to have been able to interest the artist and the architect in this conversation and thank both of them cordially for not only having found time to sit down together to talk about their work, but for their ongoing involvement during editing.

Very special thanks go to Philip Ursprung, first for his excellent moderation and secondly for editing the text, to Sylvia Bandi for transcribing it, to Catherine Schelbert for finalising the English version and to Salome Schnetz for the production management. Christine Binswanger, partner of Herzog & de Meuron, gave us an informative tour of the Rehab Centre in Basel prior to the conversation. We also wish especially to thank Eckhard Schneider of the Kunsthaus Bregenz for his active and generous support of this book series, as well as our sponsors Holcim Ltd., Zellweger Luwa Ltd. and Stiftung Erna und Curt Burgauer for their support.

Cristina Bechtler
Küsnacht, March 2004

Pictures of Architecture — Architecture of Pictures

Jeff Wall, Jacques Herzog, Philip Ursprung and Cristina Bechtler in Conversation

Basel, November 4th 2003

Ursprung

If I were to define a common denominator that links your art and your architecture, it would be the fact that you both produce images. When you entered the world of art and architecture around 1980, there was a growing demand for images. Wolfgang Max Faust's book *Hunger nach Bildern* [Hunger for Paintings] was popular in Germany and "Pictures" was the name of an artistic tendency in New York. However, the two of you had to go a long way to get there, and had to get rid of a general aversion to images. Can you tell me how you turned to making images and what kind of opposition you met from critics and colleagues?

Wall

I never had an aversion to images because my interests always started from some kind of affection for them, whether it was paintings or other things, like photographs. In the 1960s, I participated in that 'culture of aversion', which was identified with the radical art of the time. I learnt a lot from that encounter. The atmosphere of aversion was not negative; rather, it was a dialectic relationship where I, who was instinctively interested in the art of the past, had to go through a rigorous opposition to my own instincts. My pictures never emerged from a simple sense of return to the past. I see them rather as a consequence of my experience of investigating whether what we call a picture in the West still had any life to it. What

Herzog & de Meuron, Lego House, contribution to the exhibition "L'architecture est un jeu ... magnifique", Centre Georges Pompidou, Paris, July 10th to August 26th 1985, film still

was great for me about the 1960s and 1970s was that, by trying to negate that validity of the picture, whether through conceptual art or any of those associated forms, I found ways to experience my affection again, but not in an old-fashioned way.

Herzog

One great difference is that Jeff can produce images — and is expected to produce them — more than architects would be. But the question does touch on something very important, namely that we've always based our work on 'images'. We started a bit later than Jeff did, the first building was erected in 1979, and our first project was in 1978. At the very beginning as very young architects, postmodernism was emerging, something with which we've never been able to identify. And when deconstructivism came along two or three years later, we couldn't warm to that approach either. We experienced it as a field that wasn't ours. So we had to find our own way.

But we were too young and new on the market to get commissions and therefore couldn't produce buildings — technically speaking — as easily as an artist can produce images. While looking for alternatives, we came across video, which nobody was using in architecture at the time. Video images are interesting because they relate to real life. As in photography, their pictorial reality expresses things and acts that look real, so suddenly we found ourselves with a tool that would allow us to express our ideas on architecture in a contemporary form even without a concrete commission — and much more successfully than by using classical means of representation like models, plans and drawings. So we produced images of what could become architecture. (fig. p 12) The pictures showed basically traditional interiors and ordinary life. Like the movies that we love by Powell & Pressburger or Hitchcock, we wanted to develop new possibilities for (architectural) events out of things that are very familiar rather than immediately introducing a new idiom. It seemed much more interesting and subversive to us. For that reason we were accused of opting for convention and tradition in opposition to the avant-garde, for example deconstructivism and other trends that focused on audacious and uncharted ideas. Actually, the deconstructivist emphasis on newness at all costs just

Jeff Wall, *Dominus Estates Vineyard, Yountville, CA, Winery building by Herzog & de Meuron, Basel,* 1999, black and white photograph, 209 x 267 cm

bored us. It seemed like a rehashing of Russian constructivism and neo-expressionism, and we felt that a lot of potential for our generation lay in rejecting Modernism's almost ideological emphasis on novelty. Paradoxically, we focused on familiar, popular and sometimes banal pictures in order to destroy or at least avoid the imagery of the architectural zeitgeist in those days. Our beginnings, with all the images we came up with, were essentially iconoclastic.

Wall

I haven't seen very many of your buildings, maybe five or six, but, aside from the imagery, I am most interested in your use of materials. They are often unusual, and used in unexpected ways, like at the Dominus Winery. (fig. p 16) I don't see that as a programmatic project, but as something intuitive and spontaneous. If one uses terms such as 'intuitive' and 'spontaneous' one finds oneself in an aesthetic world that is very different from the kind of postmodernism you are talking about. To be intuitive is sort of romantic. I don't mean romantic in a historical sense, I just mean 'lyrical'. And to me, that is an aesthetic attitude that is connected to nature. I know that my sense of nature has always had a lot to do with the way things look, the volumes, colours, or textures of a certain object or thing. (fig. p 14) This old and probably conventional sense of nature was one of those things that were apparently devalued in discourses like deconstructivism. But I see a strong touch of this feeling for nature in your architecture. I wonder whether that instinctive quality has something to do with an acceptance of the past. Not an uncritical acceptance, but an instinctive acceptance.

Herzog

You've mentioned several things: romanticism, nature, instinct, the past. These concepts are all somehow related, for one thing because they are somewhat disparagingly considered anti-intellectual. But that also shows their potential: there is no reason to pit intellect against instinct. I've always been fascinated with the way Novalis and Goethe interpret romanticism. There is something both naive and marvellous about Goethe's drawings of cloud formations because they testify to the attempt to understand

Herzog & de Meuron, Dominus Winery, Yountville, Napa Valley, California, USA, 1998

and keep a record of nature, to assimilate it. 'Natural' means the existing physical world to us — similar to the way you describe it for yourself — and our architecture is a response to that. That's why using all five senses to explore the physical world is central to everything we do. We insist on involving them all because that approach strengthens architecture with respect to other media like movies or television, which are incredibly seductive but appeal mainly to the sense of sight. In other words, architecture is an old, archaic medium not because it has been around for such a long time but because it makes such great demands on people, drawing them in, abusing and seducing them with all the senses. Only in that way, with all this archaic variety is architecture capable of surviving and worth preserving. Otherwise it's decoration, a purely technical structure for our diverse activities.

Ursprung

Materials can indicate the passing of time or time as a process. For both of you, 'the new' has always played a less important role than the 'transformation' of something pre-existing. Can you tell me something about your interest in historical time?

Wall

Hannah Arendt suggested that artistic creation produces things that are meant to last a long time. This means that the materials out of which sculptures or buildings are made are supposed to get old. I was in Sicily last week, looking at very old things. In the Mediterranean area you see a lot of structures that are aged, not necessarily ruins, but old. And their ageing process is extremely important for their beauty. Once you get interested in materials and in the ageing of material, your relationship to the past has to be affected somehow.

Herzog

Anyone who thinks up, creates, produces or just uses objects — everybody, that means everybody — is confronted with the fact that things change, wear out, erode, decay or somehow disappear. That also applies to architects and their buildings, except that the scale is larger. It's interesting,

Herzog & de Meuron, Eberswalde Technical School Library, Eberswalde, Germany, 1999

though, that there are very different ideas about how to ensure that things and materials have a longer (physical) life and longer (spiritual) survival. Vitruvius's *firmitas* best reflects the western bunker mentality, but I prefer *venustas* — you can come across fragile materials and forms of construction, especially in Asian cultures, that manifest an extraordinary resistance and life span despite their fragility. Palaces and temples made of paper, bamboo or wood sometimes last longer than ones made of stone.

Wall

Like the Japanese concept, *wabi sabi*. It includes the idea that fragile things survive because of the way people relate to them, and that their fragility and decay are an important part of why we relate to them the way we do.

Bechtler

The concept of ongoing renewal can also be a ritualised procedure, like the Shinto shrine in Ise, which is dismantled and completely rebuilt every 20 years — and it's been done for the past 1300 years! The idea of the ephemeral is more rooted in Asian societies while the Western world seems to focus more on conservation.

Herzog

Survival works because people love and admire a building or are simply enthralled by its beauty. We are extremely interested in this idea of beauty and it is, of course, associated with the concepts we were talking about before — nature, romanticism, etc. We can only hope that all of our fragile constructions, like silk-screened panels, glass panels, moss, wood or walls of stones piled up without any mortar, will generate enough magic to encourage the love and devotion necessary for proper maintenance. (fig. p 18)

Ursprung

There seems to be a fundamental difference in attitude here between the two of you. I was interested in the fact that you, Jeff, claim to make your black-and-white pictures last for a hundred years …

Wall

… a thousand years … (laughter)

Ursprung

… whereas Jacques says that he is fascinated by fashion or video because they are ephemeral.

Herzog

The two approaches are not mutually exclusive …

Wall

… because ephemeral things can last a long time. If one cares about them, they are going to survive, even in a decayed state.

Ursprung

Although you agree, I'd still like to explore where your attitudes diverge. Jacques seems to be horrified by the very idea that someone could define or pin down his ideas. His rhetoric is basically one of negation. Jeff, on the other hand, seems completely open to categorical definitions. He 'fits' perfectly well into discursive practices and his rhetoric is that of affirmation. I often have the impression that, for Jacques, the world is full of meaning and images, and that he is trying to create some 'empty' space. On the other hand, I have the impression that, for Jeff, the world is a priori empty and waiting to be filled with images. Am I correct?

Herzog

Actually, to us the world is full of meaning and also full of images and full of a priori's. We just have to deal with it. Protestantism, the dominant cultural heritage in Switzerland, produced the antidote of iconoclasm, whose contemporary form in art and architecture is the empty space. Emptiness, the absence of images, always seems laboured and contrived and reminiscent of religious fervour. We're more interested in destroying pictures in

fig. p 20: Jeff Wall, *No*, 1983, transparency in lightbox, 330 x 229 cm

Herzog & de Meuron, Signal Box, Auf dem Wolf, Basel, Switzerland, 1994

order to make room for other ones. Destroying is more interesting than preventing or suppressing.

Wall

When I come to Europe I notice differences from when I am home in America. (By 'America' I mean both the United States of America and Canada.) Here in Europe, older ways of doing things and older kinds of spaces and structures are really palpable — such as the house we are in this afternoon. Generations have passed, looking through that window in front of us. In Europe I sometimes feel an almost suffocating sense of the past. In America you don't have that feeling, even though America is now over 400 years old.
At home, we can still have the experience of the pathless, the less mapped, even though our places are also getting old and very settled.

Herzog

In Switzerland you can't experience anything that isn't 'mapped'. All the trails are laid, named and sign-posted. Everywhere there are railroad tracks, streets, houses or signs telling you not to do something. Switzerland is totally urbanised and yet not truly urban. Switzerland is totally modern and still caught up in the obsolete image of being a natural environment that contains urban areas. This wishful self-image is so ingrained and so successfully communicated to the world outside that even people who occasionally visit Switzerland believe in it. But in reality Switzerland, possibly even more so than Holland, is a topography of artificial urban development, sprinkled with undeveloped land.

Wall

I wonder if what we call modernity isn't really very old. The new keeps happening and I feel that we can find what we call 'the new' a very long way back into the past. This has to do with my sense about why the arts of the past remain in some critical way available to us. 'The new' has happened before.

Herzog & de Meuron, REHAB Basel, Centre for Spinal Cord and Brain Injuries, Basel, Switzerland, 2002

Ursprung

When I went to see you in Vancouver, Jeff, I found it full of traces of the past. I felt that I could actually see through the layers of the past, see history. In Europe, and particularly here in Switzerland, I have a problem seeing history because everything is constantly being fixed and varnished and polished.

Wall

I rarely have visitors in Vancouver, but when I do have them and show them around the city, I sometimes feel that it is disappointing for them. Sometimes I even feel a little ashamed because Vancouver is really not an impressive city, as a city. It is very ordinary, even worse than ordinary. But that's what it is, and I'm afraid that is what modernity is like when it is fresh. We often react negatively to that 'worse-than-ordinary'. In my pictures I try to perceive it as the actual environment in which we live, as the result of all our labours and errors. I think that's an important way of looking at it and I know that architects sometimes look at it in that way too, partly out of their sense of trying to 'learn from Las Vegas', to learn from the vernacular. One of the essential things about the vernacular is that it is unimpressive, it is ordinary, worse than ordinary. It is the essential phenomenon of what we call 'the new'.

Herzog

Last night, I was studying your famous picture *No* and trying to figure out what type of city it shows. (fig. p 20) You can't see the sky, only the city, any American city, or what you could call a modern city 'when it is fresh'. It's a city without all those specific features that inevitably emerge over the years in the process of acquiring a history. In your pictures you seem to go for an anonymous context that does not refer to any specific city.

Wall

It is very important not to use a city as a specific background. Sometimes I find it very boring to find out that in one particular town there is a very good kind of ham and in another town nearby there is another very good kind of ham and that you must go to this town or region to taste this ham,

Herzog & de Meuron, Forum 2004, Building and Plaza, Barcelona, Spain, 2004

this wine or this rucola and so on. It gets tiring realising that every place is so special, has its own special thing — in the end it is all ham, wine, rucola. You can see why people like Coca-Cola because they don't need to have this 'special experience' all the time.

I don't like the idea of the 'genius loci', the special spirit of a place. I don't like the idea that everywhere people feel they live in a very special place that is precious to them, and they are anxious for all the world to know how special it is. I don't deny that there is some truth to all this, because I like this ham better than that one, too. But as an artist I don't want to feel that way; I want to feel detached from 'the local' whilst being in it, not valuing it, just trying to observe it.

Inevitably you are attached to your own place, and it's okay if it has a personal value to you as an individual, but I don't think it's good as an artistic value.

I dislike the notion of 'the local'. To me, it means parochialism and provincial power structures. I like the idea that great communities can be formed by people who insist that they are not special, not somehow selected specially by heaven because of where they live. I want to emphasise that we need to value precisely what is *not* special about us. That way, we are less inclined to create local cults against 'outsiders'. I don't want to present any city as a place more special than any other place.

Herzog

That is something art can achieve. Art can express the idea of something that is "not special", in other words, of general validity. Cities can't do that although it is a goal that urbanists keep trying to achieve: a generally applicable urban form with no specific idiosyncrasies and yet still competitive and self-confident. You can see efforts in that direction, for instance, in the geometrical layout of Roman cities, in Haussmann's boulevards in Paris, in the modernist proposals of Hilbersheimer or Le Corbusier, in socialist cities built after the war and now in the new cities that are being built by the Chinese. There is always the eternal pipe dream of emancipation because as cities age they become increasingly specific, both to their advantage and disadvantage. The advantage is that residents generally want to make their city better, more beautiful, stronger or just simply different and more

Jeff Wall, *The Stumbling Block*, 1991, transparency in lightbox, 229 x 337 cm

special than other places. Cities are a kind of battle zone; they physically reflect human strengths and weaknesses. They are like a petrified psychogram. They can't help becoming specific, which leads to fascinating, spectacular phenomena like superb buildings, squares and facilities, but the emphasis on difference at all costs and certain affectations are sometimes pushed to almost ludicrous extremes.

Wall

If you're an architect and want to make a building in a city, you want to make a good building. And by making a good building, you are creating something that people want to appreciate and when they appreciate it, they feel they are in a good place, which makes that place better than another place. I want to suspend that. And therefore to photograph a place like the street in *No* is as if all the ambitions that have come together there have been both realised and lost. And what is left over is the experienced world that contains both successes and failures. I don't want to photograph catastrophically failed spaces, partly because that has become a sentimental genre, and also because it is not really that interesting artistically. What is interesting is this zone in which attempts to create a world are in process, have not been completed, probably cannot be completed, and at the same time are being encountered, affected by other forces. So there is some element of failure, there is some element of success, and there is some element of indifference. All those blend to create the picture's atmosphere. This is the photographer's good fortune — we need neither urban success nor urban failure. As an architect, you don't have that freedom.

Herzog

Plus, as a photographer, you are able to produce more than one copy, while we mostly make one-of-a-kind products. Prefabrication and repetition, which apart from the economic factors also embody the utopian attempt to design a generally valid urban form, have all but disappeared from the building industry. Originally we planned to build signal boxes clad in copper, like the two we built in Basel, in all of Switzerland's major cities. (fig. p 22) It would have been interesting to be able to compare the same or very similar buildings in various local contexts and to set up a kind of wide-ranging urban territory.

Wall

I don't agree, because a designed environment only presumes to be completed. But it is never experienced entirely that way. For example, the Rehab Centre we visited this afternoon is very much designed and well designed, and the environment is strongly shaped, but events will always be taking place there for which you could never plan. (fig. p 24) That's just the way things are. You told us, Jacques, that you were making videos to imagine an interior life in your project before it exists. But you can't really know everything about that interior life, and that's what's interesting about buildings. As they get older such unexpected, unanticipated patterns show up. I am sure it must be upsetting for an architect to go to a building he designed, and discover that the people don't use it the way he had in mind. I am sure these things happen, and they make the building more interesting. That's what photographers like. Even the most carefully designed environment can never really be concluded. I suppose that the most sophisticated designs don't want to get concluded.

Ursprung

Many of Jeff's pictures look like they were film stills. Jacques, you commented on the importance of film for the work of Herzog & de Meuron. What films are interesting to you, and how do they relate to your work?

Herzog

In the films of Powell & Pressburger, Hitchcock or Antonioni, it is fascinating to observe how architecture is treated like a character in the plot. For example the vertiginous mansion in Hitchcock's *North by Northwest*, the gloomy interiors in Powell's *Small Back Room* or the winding stairs in Antonioni's *Identificazione di una donna*. What appealed to us was the idea of ordinariness, ordinary architecture, people wearing ordinary clothes, which expresses the whole drama of human life much more subversively and profoundly than any over-expressive, fictional stage architecture. Following that track, we tried at the time to come up with an architectural vocabulary that was inconspicuous and 'normal'. Creating something inconspicuous was, of course, much more difficult for architecture in those days, than producing something new. That is why we borrowed the aesthetics of a new medium, not yet established in the field of architecture, and produced video stills of our designs with real people in them in order to draw attention to our work, its potential, and ultimately ourselves. We had hardly any commissions and so we composed these images and imagined how people would move around in them. Today, with all the projects that we've done, we can go to the buildings themselves in order to observe, photograph or otherwise document how people circulate in them.

Wall

How do you influence what will be there?

Herzog

Architecture can facilitate life, make it enjoyable and inviting, or conversely, make it more difficult. Sometimes — and ideally — the architect succeeds in offering a potential that goes beyond the functional implementation of a brief. We're trying to do that at the moment with the Forum 2004 project in Barcelona. (fig. p 26) We've practically reversed the brief and are "misusing" the actual program (a huge exhibition space and auditorium), in order to generate space that can be used by the general public apart from conferences, congresses and exhibitions. We want the space to be for normal life and normal people in the neighbourhood, not just for specialists coming into the neighbourhood from outside and not related to it. We raised the entire building off the ground so that there would be a huge covered plaza underneath and convinced the mayor to set up a market and a chapel there. We'll see whether it works. In any case, the idea is to make room for urban life to settle down there in order to counteract the threat of deterioration in the area.

Wall

That's another reason why I think buildings should age, because their use could evolve over time. The cinema is very good at depicting the

Jeff Wall, *The Crooked Path*, 1991, transparency in lightbox, 119 x 149 cm

emergence of the identities of spaces by the way people inhabit them. This is the case especially with the neo-realists and then Antonioni, perhaps because they live and work in Italy where buildings are constantly being reused. So it is almost as if the cinema deals with the 'after-use', the second or third use of a building.

Photography does that too, in a less dramatic but maybe a more permanent, more contemplative way.

Ursprung

Is that why you build like to build sets for your photography?

Wall

No, I only build sets if I have to, for practical reasons, and occasionally because I want an artificial environment.

Herzog

How did you shoot *The Stumbling Block*? (fig. p 28)

Wall

I photographed the background, along with a number of the figures, out in the street. Then I took measurements of the foreground sidewalk area and reconstructed the first 20 feet of it in the studio. I photographed a number of the main figures in the studio and then combined their photographs with others taken in the street, using the computer.

Herzog

Did you project the city onto the background?

Wall

No, that sort of projection is completely unnecessary now.

Herzog

Was it just in your imagination?

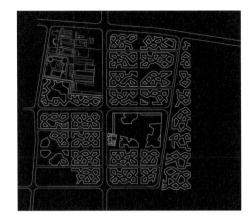

Herzog & de Meuron, Tree Village Campus, Beijing, China, masterplan 2003–2004

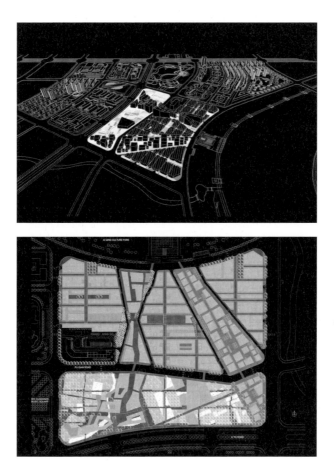

Herzog & de Meuron, Jidong New Development Area, Jinhua Commercial Cultural
and Entertainment Center, Jinhua, China, project 2003–2004

Jeff Wall, *Outburst*, 1989, transparency in lightbox, 229 x 312 cm

Jeff Wall, *Mimic*, 1982, transparency in lightbox, 198 x 229 cm

Jeff Wall, *Restoration*, 1993, transparency in lightbox, 119 x 490 cm

Jeff Wall, *Dead Troops Talk (A Vision After an Ambush of a Red Army Patrol Near Moqor,*

Afghanistan, Winter 1986), 1992, transparency in lightbox, 229 x 417 cm

Wall

No, all the pictures were made either at the actual location on the street or in the studio, and were then combined in the computer.

Herzog

Do you ever feel the need to rearrange the architectural background — for example, take away one building and put another one there, higher or lower, with more or less glass, in order to get an 'ideal city'?

Wall

No. I like accidents and odd incomplete situations because that's in the nature of photography. I think an architect can imagine a project as a possible, or provisional 'complete situation'. I guess as an architect you can have a sense that this completion in your work is possible, whereas I always encounter incompletion. Not that I am looking for it, but because I feel that it is the way we actually live with buildings. So the incompletion and the subliminal elements of what I might call 'the building to come' are always hovering in the sensibility of those who are paying attention to the experience of the space. I think that is also true for natural space.

Herzog

The idea of incompletion is relevant to architecture in two respects: through physical wear, ageing, and through the unpredictable and changing way that people use architecture. A lot of our projects incorporate incompletion as an integral part of the design. You might see it as an alternative strategy to assertions of permanence or *firmitas*, as illustrated for example by various neo-classical approaches to architecture. But whether or not the projects are better protected in that way from demolition, destruction or being forgotten is a moot question.

Wall

I think you try to speed up the process by which the building's use starts to mutate. You try to find ways to accelerate the historical process. I don't think that you can succeed, but it doesn't matter because what it really changes is the nature of how you practice architecture.

Herzog

That would mean that we envision the process of transforming architecture and the city in order to transform ourselves? We have described our architecture as an instrument of perception in order to understand life.

Wall

I think that is very modernist. I think in pre-modern culture, the manifest use of a building was determined in a different way. A building that was built as a church was used as a church and there was little question about that. One did not have the right to suggest other uses. But this probably applies only to prominent things like churches. For lesser structures, and in the villages and smaller, less central towns, with their beautiful crooked paths and streets, people could renegotiate space, rethink it and experience it as an everyday thing.

Bechtler

Like your picture *The Crooked Path.* (fig. p 32)

Wall

Yes, I tried to express something of all this in that picture.
Nowadays we tend to dislike the idea of anyone dictating the ultimate way to occupy a place or a space, the ultimate way to use a building. It's not considered very democratic. At the same time, the architect, by determining the manner of occupation, becomes a new, modern kind of authority figure, since now he wants to let others discover their own path as well. He wants others to alter, maybe even violate, his original plan. And that suggests that he is making a different kind of plan now.

Herzog

Architecture and city planning cannot escape the dilemma between determination and freedom. We are currently doing urban planning for large sections of Beijing and Jinhua, and we're trying to find new answers to this problem with a labyrinthine topology of space, with what you might call a multiple-choice topology. (fig. p 34, 35) The neighbourhoods are being built so quickly that we'll have the opportunity not just to see the finished

Herzog & de Meuron, Ricola Storage Building, Laufen, Switzerland, 1987

product but also to observe the process of transformation as the years pass — that's the most exciting prospect we've ever had in doing a project!

Ursprung

That is a crucial point. You enlarge your surface the way an animal enlarges its silhouette in self-defence. In a way that's a very experimental means of defending yourself against time, since you don't know what is going to happen with the buildings and perhaps taste is going change anyway in a couple of years. So you have to speed up observation, reaction, discourse. You have to set up traps, like de-automating perception, to provoke discourse. This might relate to your practice to write a lot, Jeff. You cannot take every viewer around your photographs but you can guide your readers in using discourse. Through art criticism you can also speed up the process of reception and validation. In fact, you are trained as an art historian and your professional practice followed two paths for many years since you worked simultaneously as an artist and as a historian/critic. And Jacques, you trained as an architect, but worked as an artist until the mid 1980s and, like Jeff, you've done a lot of writing, too. Can you tell me more about this dual practice? Are your writings commentaries on your own art and architecture, or are they autonomous works of theory? Why do you both write much less today?

Herzog

To me, writing has always been a tool that I use in order to understand. Sometimes a door opens up onto vistas that I haven't seen or thought of before. But there's also the danger of developing your thinking too much in the form of writing so that you rely too much on the logic of the written thoughts. I'm convinced that only literary, poetic writing really makes sense and can pass the test of time. When architects write, it's a technical or journalistic writing for the sake of personal marketing. I want to restrict myself in that respect because it doesn't make any sense and it takes too much time.

Wall

In my case, there is no particular reason for writing less. Writing is difficult, especially when, like me, you have no literary talent, only some ideas to talk about. I wrote quite a bit during the years I earned a living teaching, but nothing of that was ever made ready for publication, and so it was never published. Then I wrote some critical essays because other artists invited me to write something for their exhibition catalogues. I probably took that work more seriously than I should have, and wrote more than I should have, and more intensely than I should have. Maybe that's why the invitations stopped coming.

Herzog

Are you perhaps unconsciously trying to lead the viewer, the consumer, the person who looks at your work in a certain direction?

Wall

I think I have made pictures that had that quality, that seem to have a specific meaning. I am thinking of pictures I did, mostly in the 80s, such as *Outburst*, the picture of the angry factory manager. (fig. p 36) In pictures like that I did want to create access to what I might have thought to be the meaning, so people would have a direct, immediate experience of a situation whose meaning they might appreciate. That tends to suggest a dramatic picture, in which much is made visible. Slowly, I have come to feel that meaning is almost completely unimportant. Because of our education in art over the last 20 or 30 years, people expect to relate to art by understanding it, by apprehending what it means. But I may have returned to an older, simpler point of view, namely that we don't need to understand art, we need only to fully experience it. Then we will live with the consequences of that experience, or those experiences, and that will be how art affects us.

In pictures like *Outburst*, I was experimenting with that sense or value of direct meaning. It is interesting to experiment with something that you may think is aesthetically inferior. Often you don't know the nature of the rule you are following until you attempt to break it or to rewrite it. So I don't want to say that it's not a good way to make pictures, it's just not the way that interests me at the moment.

Herzog

Could you describe in which way you see meaning to be more represented in this particular picture than in other ones? I see it as very expressive the way this guy is screaming at a garment worker but there's no difference to me between that and the gestural impact of a work like *Mimic*. (fig. p 37)

Wall

You're probably right. I suppose what I see as its expressivity lies in the intensity of the gestural language and of capturing the gesture at what seems to be its culminating moment, a moment that seems to reveal something basic or essential about the situation. It isn't different in kind than other pictures, but it seems to me to be taking all that gestural work further than some, if not many, others.

Herzog

How are you going to eliminate meaning? I can certainly understand your idea of the viewer simply fully experiencing a picture. But you can't ignore content and that's probably not your objective anyway in view of the failure to do so in abstract painting or minimal art. What strategies or concepts do you pursue with regard to meaning? Some of your works, like *Restoration* or *Dead Troops Talk (A Vision after an Ambush of a Red Army Patrol near Moqor, Afghanistan, Winter 1986)* remind me a little of the Byzantine tradition of inverted perspective where the simultaneous and equivalent presence of various scenes in the same picture communicates an 'impossible' spatiality, a kind of discontinuity that forces viewers to move back and forth and to look at the picture from different distances. (fig. p 38/39, 40/41) The fact that the viewer has to move around in itself prevents restricting the focus to only one meaning and adds a certain element of chance.

Wall

The accidental, the contingent, is what makes photography exciting. Any aesthetic of photography contains this notion of contingency, and so of course, in a certain sense, so-called overly designed photography might be considered less photographic. I have experimented with this overly designed photography and don't think it is 'less photographic'…

Herzog

… it's more cinematographic.

Wall

Yes, probably more cinematographic and less photographic in the classical sense of photography. The point is that the multiple elements you're interested in are usually created through accidents, through chance. It is fascinating that any scene that is put in a frame contains accidental combinations of elements. Here in this room at the moment, certain objects happen to be placed on the table. They are not necessary for our meeting here, there is no call for those things to be here, they are not required. So in that way, they are accidentally present. If I should photograph us here this afternoon, these objects would be a contingent element in the picture. Nearly every combination of elements is accidental to some extent. Sometimes I stand on the street corner (this is a photographer's game) and try to predict what the next person who is coming around the corner is going to look like. You will likely never guess right, and that unexpectedness is fundamental to the medium. So if you let that happen in one way or the other, be it through artifice or through lack of artifice, in whatever technique, it will give rise to multiple shapes, colour combinations, accidental figurations and other marks of contingency. All of those things are essentially dissonant. They're not harmonic, in the sense that a painting by Poussin or Cézanne can be harmonic. Photography doesn't have that harmony unless you impose it by means of composition.

Herzog

I never think of the accidental when I look at your pictures although it is, of course, extremely important in photography and in life as well. Maybe that's what unconsciously appeals to your viewers, this aspect of chance. Chance, or rather being open to unanticipated or different kinds of change, is also our objective in architecture and in the urban projects in China that I was talking about before.

Harmony or overdetermination is a problem that applies not only to architecture but to art as well. As I said, it's fascinating the way Byzantine artists distorted perspective in order to make viewers actually move about in front of their pictures and to avoid nailing down a specific focus. Later, painters like Delacroix or Tintoretto — you've referred to them in your work, too — tried to create the same effect using different means, like a sense of scale. Do you see this contingency factor or something photographic in their work?

Wall

I've said many times that I think one of the great qualities of western painting since the Renaissance is its sense of scale, that scale which is related to our body and to the room in which we look at the picture. Scale helps us to realise our own mobility and its role in our experience of works of art. There is great pleasure in that mobility. Great baroque and post-baroque art has this fantastic combination of scale and order that makes you want to move back and forth to see it at different distances. It immediately suggests that there is no single way to see a picture.

Herzog

The strategy of offering viewers different levels of reality and scale at varying distances applies to the medium of architecture as well. We worked with that idea when we built the Ricola storage building in Laufen, under the influence, at the time, of conceptual and minimal art. (fig. p 44) When you move up very close to the building, the image of the façade falls apart: you only see single components, nothing but paltry screws and planks. It was an important step for us, a new way of looking at the surface of the building, investing it with a kind of materiality and reality of its own. I think surface, as such, is not a major concern in your work.

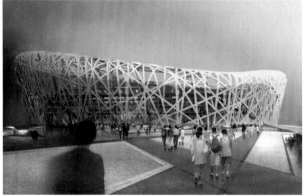

Herzog & de Meuron, National Stadium, The Main Stadium for the 2008 Olympic Games,

Beijing, China, completion projected for 2007

Jeff Wall, *The Destroyed Room*, 1978, transparency in lightbox, 159 x 234 cm

Wall

No, no, I am very interested in it. But I want my surfaces to look like they do, with very fine grain. That is almost the only thing I am interested in — things like the qualities of the surface, of the grain, of the physical, formal aspects of depiction.

Herzog

But you don't treat it as a self-contained issue, like for example Gerhard Richter or, increasingly, Thomas Ruff in his recent work on jpg structures …

Wall

… You mean making pixels visible. I don't want visible grain or pixelation, I find it distracts from the pictorial qualities I think are more important. The crucial quality of photography is the sense that the picture surface is invisible, or seems invisible. This distinguishes it from painting; with painting the surface is physically visible and then the illusion springs from that visibility. When photographers try to imitate that, they almost always fail, because photography is not the same medium as painting. It is the invisibility of the surface that is so dramatic. What you are experiencing in photographs, in the moment of seeing the thing depicted in the photograph, is the invisibility of the surface.

Herzog

Is that why you illuminate it, to make it even more immaterial?

Wall

It can, although I don't think that's essential. I see the emphasis on the mark of the pixel as indicating the imitation of the norms of painting. I feel that to emphasise the invisibility of a photographic surface, its intangibility, is truer to the medium of photography. Overemphasis on grain is a way of making photography seem painterly. If the grain in your photograph is too big, it is because you've enlarged the negative too much. You're out of scale.

Herzog & de Meuron, Blue House, Oberwil, Switzerland, 1980

Herzog

Then are you essentially interested only in the picture as a picture?

Wall

Yes, in a way, but there is no such thing as a picture only as a picture, because the being of a picture is a complex product. That would be like saying a building is only a building. It's not so simple.

Herzog

I think it's not quite the same because the picture has a philosophical tradition that goes back to Plato and the idea of the icon, whereas the building has always been described in its material context, in its functional, concrete context.

Ursprung

The architecture of Herzog & de Meuron, like the Eberswalde Library (1999), is often criticised for being overly focused on the façade. (fig. p 18) Physical space seems to be less important than the image transported by the façade. There seems to be a specific kind of spatiality, a spatiality that is different from the 'sculptural' space one can observe in, say, the architecture of Mies. Jeff has photographed both architectures, your Dominus Winery and Mies's Barcelona Pavilion. (fig. p 14, 16, 58/59) Which is more 'photogenic'? Can you describe this difference in the treatment of space? Is Herzog & de Meuron's emphasis on surface related to what Jeff calls the "invisibility of the surface"?

Herzog

We've focused so intensely on the surfaces of buildings because it seems to us such an interesting and neglected field of activity. In some buildings, like the Dominus Winery or Eberswalde Library, the outcome has been especially spectacular so that some critics, especially those who have not actually seen the buildings themselves, tend to reduce our work to the treatment of surfaces. Our buildings are designed as 1:1 architecture and are therefore always better on site, no matter how good they might look in a picture.

Wall

I think the picture as we call it in the West has been misrecognised almost continually, in other words, people — whether they are learned or not — tend to enjoy misrecognising a picture as just a picture. It seems to be the most natural thing in the world and I think the reason for it is that the western type of picture, which you might say culminates in the photograph, resembles the way the world looks when we are not looking at pictures. To produce that effect, you need great skills, especially in drawing and painting. It's so difficult to achieve but seems to be natural when it's done. That illusion is central to the pleasure, which is why we always misrecognise the picture. That's probably not true of a building. We can recognise architecture as what it is although we might misrecognise its latent uses.

Bechtler

Is that why you try to avoid meaning? Or just express it in a psychological sense but without the connotation of a certain society or narration or symbols?

Wall

Yes, I don't have any interest in those things because I think the picture as a picture is something that we experience with our whole self — if we have the capacity to do that — some do, some don't — and it affects us, and we don't really need to know how it affects us. The pleasure is what affects us, it's the enjoyment that matters to me, it's the only thing that matters. I think that's simpler than architecture. Jacques, do you consider architecture as an applied art in a traditional sense? I ask this because if there is a practical application, then enjoyment cannot be the only thing that matters.

Herzog

Obviously architecture is not a fine art. There are constraints, like budgets, briefs, etc. But that's not the problem. The problem for architecture today is not a lack of freedom but freedom! Traditions and architectural typologies have become obsolete; there are no rules and no directives anymore on how to build a church or a museum or a city. But it's thanks to such rules that cities and architectures of the past from all of the world's traditional civilisations look beautiful to us. Paradoxically, beauty did not become an architectural problem until the freedom of industrial and modern architects and urbanists overshadowed the lack of freedom imposed by tradition. The rise of this kind of freedom calls for a new type of architecture with an aptitude for artistic and conceptual thinking because the problems facing architecture are becoming more and more related to those of the fine arts. That freedom has produced some very fine edifices and caused a burst of global competition, but the downside is the construction of some of the most abominable buildings in history. Sadly, the latter grossly outnumber the former.

Wall

That's why I said that, with a work of architecture, enjoyment is not sufficient; you have to use it, to relate to it practically, in some way.

Herzog

Yes, because — as I said — you have to deal with a brief, a programme, a budget, etc. But that's not the real burden anymore. The real burden is being faced with a white sheet of paper and having to develop your own world. And that resembles the artist's job, doesn't it?

Wall

Good architects must have a talent for bearing that burden, so that the programme or what the clients want doesn't feel like a burden. Talent for architecture must involve not seeing that as a problem.

Herzog

Ideally, yes.

Bechtler

You might say, as you see it, that you are faced with a tabula rasa, but I think that art and architecture can also be a kind of seismograph for questions concerning society. There is a philosophical meaning beyond aesthetics. For

Jeff Wall, *Morning Cleaning, Mies van der Rohe Foundation, Barcelona*, 1999, transparency in lightbox, 206 x 370 cm

example when you design something like the National Stadium in Beijing, a much larger context is involved — cultural and social issues and their connotations, the actual environs, lighting or the acceptance of materials. And then there's also the process of formulating these concerns.

Herzog

In our projects in China, and most especially the National Stadium, heralded by the media as the national monument of present-day China, we clearly have to address the relationship of contemporary architecture to tradition. (fig. p 50, 51) Traditional images and codes, like feng shui, are much more deeply rooted in the collective conscious in China than comparable traditions anywhere else in the world. We couldn't possibly fruitfully incorporate the way that works without the help and co-operation of Chinese artists, especially Ai Wei Wei. And, of course, that complements and enriches our work rather than being a burden. A great deal has come to light about which we actually knew nothing at all. The real burden is our own blindness and the challenge is how to overcome it, in the same way that artists are always trying to pinpoint their unconscious assumptions in order to overcome them. Although I wouldn't claim that architecture is art.

Wall

A building is art. I wouldn't suggest it isn't art. It's just that there is an old distinction between the arts that don't have to do anything but please, like painting, and then this curious other art that can be experienced by people who take no pleasure in it, for example, the user of a building who goes there every day and never appreciates its forms or colours. I am sure you have met people who use your buildings in that way, and yet their use is legitimate and has meaning. A person who takes no joy in the use of your work can still enter into an authentic relationship to it, whereas a person who takes no pleasure in the forms of a painting or a photograph develops no legitimate relationship to it.

Ursprung

And I suppose part of that relationship has to do with meaning. You entered the art world by explicitly offering to produce meaning, like titling your works. For a long time, people were calling their work 'untitled'; content was not allowed. Starting with *Destroyed Room*, you established a language of your own. (fig. p 52)

Wall

I had thought of taking all the titles off my pictures, but I guess I can't do it now.

Herzog

Why would you want to do that?

Wall

Firstly, because slowly I began to feel that any attempt to point out the meaning of the picture tended to inhibit the experiential side, to not let the picture be what it is. I have also come very much to believe that I don't really know what I am doing, in a programmatic sense, and I am more and more pleased with that. I know what I'm doing technically and practically, but that's different. A title sometimes feels like it is a statement that you know what you are doing.

Bechtler

Yes, and it can also have the effect of channelling perception.

Herzog

I don't quite agree. A title can be great and memorable if it really complements or enhances the way you appreciate or approach something. Sometimes a title is superfluous. When I asked you about *The Stumbling Block* before, I remembered the title only because I'd read it again yesterday although I certainly hadn't forgotten the picture. (fig. p 28) I still had a perfectly clear image of it in my mind. So, in this case, it seems to me that the title doesn't make a compelling contribution; it doesn't add to the complexity of the picture. But, ultimately, that's what titles should do. We gave names to our very early projects, like the "Blue House", as a kind of overdetermination, at a time when nobody was doing that. (fig. p 54)

Wall

What do you think about that now?

Herzog

I think it can still work. At some point we found it counterproductive because we wanted to move in a more abstract direction in order to leave the door wide open to all kinds of interpretation. But we're much more relaxed about that issue now. We work together in groups — maybe this is different from the way you work, Jeff — and that has an impact on how we find agreement in the process of exploring the identity of a potential or future object. Internal communications play a role in that process as well. Sometimes you can almost sense the project revealing its profile, like a photograph emerging on paper in the developing bath.

Wall

Then you start calling the project something, spontaneously.

Herzog

Exactly. And sometimes that spontaneous reaction can produce a great title because everybody agrees on it without having analysed it.

Wall

Interestingly, most people cannot remember the titles properly, which suggests that there is a discord inside the naming process. The fact that people forget or get the titles wrong interests me because it suggests that what I set out to express is not what I expressed.
I think you were saying, Philip, that I had a polemical intention back in the 70s …

Ursprung

Not polemical. To me it is fundamental that you gave names to your work. In my view, it was a way out of the dead-end of conceptual art. The same goes for your buildings, Jacques. I really like the fact that they have names even if I forget them or mix them up.

Wall

I guess that's what titles are there for … to be misremembered …

Ursprung

They may result in a certain semantic overdetermination, but there is a positive side to it because "misremembering" them, as you call it, demonstrates that you can never completely fix meaning.

Wall

That's a very interesting way of looking at it because I think a title can only be unproblematic if it's entirely generic like "Still Life with Apples". It's almost like a number or a given name like Jacques or Philip; it simply reiterates what you are seeing.
In the early 70s, it was provocative to assume that we could come back to a relationship with our pictorial tradition. I have been criticised for apparently "simply resuming" the pictorial tradition. But even if I had tried only to resume traditional forms (which I did not), that would still be an experimental gesture, just as experimental as, say, pouring paint off a stick from a coffee can, the way Pollock did.

Herzog

That reminds me of the flack we got from critics about our early works because they looked so unspectacular and didn't meet the expectations of what was supposedly avant-garde. Like Jeff, the fact that we might recur to traditional aspects, for instance, giving buildings names or titles or incorporating old-fashioned forms and materials like natural stone, does not imply a latent moralism or a penchant for convention. Quite the opposite. It's an experimental act and an attempt to shake off the ideologies of postmodernism and emerging deconstructivism. Today our designs look much more spectacular and nobody can criticise us for a lack of inventiveness or richness of form. Actually the problem is the richness itself, countless variations that flood the world of architecture and art, and generate a kind of blindness. The problem, as always, is to escape the tyranny of innovation.

Wall

Now, so many people are so good at, so skilled at, making really rather good things, well-thought-out, well-designed, well-crafted, very self-aware things — art works, buildings, publications, events, and so on — all our 'cultural production'. It isn't awkward, bad and ugly, it's good, but at the same time mediocre, chic, trivial and therefore depressing. Depression seems to be a major issue these days. Maybe that's not because of failure, as you'd expect, but because of the widespread success of mediocre culture, the elevation of the quality of the mediocre.

Herzog

Or you might say that so many interesting and beautiful things are happening at the same time that it's …

Wall

… getting depressing. (laughter) The kind of artistic freedom that we now have has both positive and negative effects. The avant-garde tradition, both in architecture and the arts, was aimed, paradoxically, at reducing artistic freedom through ideological prescription even though it was based on a principle of artistic freedom. This is what is continuing in the academies. But now that many people do not accept the strictures of avant-garde-type thought, we have gained this immense liberty. Naturally, there are some very positive things that come out of it, but also some extremely depressing results. Not bad art, as I said, but a better-than-ever mediocre art, what Catherine David called "art lite".

Herzog

The end of utopias, end of history, end of difference. Scholars like Baudrillard have described and analysed all of that. Baudrillard even laments the end of indifference. Indifference was the ultimate pleasure of intellectuals, who held up the alternative of cold indifference to a world greedily infatuated with difference. Now even that's gone down the drain. But we're still around and still making our products: pictures, buildings, etc. Though we seem to be in a situation where the impact, the meaning, the tension of whatever we do has been disconnected, like a global outage. Outages, natural catastrophes and terrorist attacks at the dawn of the 21st century have paradoxically made the world very real again. It's a tremendous challenge to come to terms with this new situation. As we see it, it could be a great chance for renewal.

Wall

I think that the situation is favourable for those who have the ability to make use of it. For somebody who has an aim, discipline and judgement, and a sense of what is legitimate in art, this is probably a very good situation in which to work because the external aesthetic and ideological constraints are so weak. I suppose we just have to take the exuberant and confident diffusion of mediocrity as one of the major symptoms or consequences of our extraordinary prosperity as a culture.

Herzog

The computer is the accomplice of this rampant diffusion of mediocrity, as you call it. We observe — with a mixture of admiration and detachment — how quickly the computer provides our teams with information and references on a new theme or a new project. Once the pictures and information come up on the computer screen, it's very hard to oppose them. That is, it's hard to distinguish the computer, as a purely technical aid, from its influence as a device that is not only very seductive but also levels everything out and therefore encourages the mediocrity you're talking about.

Wall

I see the computer and its abilities in rapid modelling and simulation as a part of *techné*. I am trying to dethrone it to a certain extent, even while I use it, or maybe, in the process of using it.
Photography always jolts you with sudden technical problems. Some of those problems were simply insoluble with the means of photography alone. But being unable to resolve them actually heightened what we could accomplish, and so they determined what photography was. For example, at the Barcelona pavilion on a sunny day it is impossible to photograph the inside and the outside at the same time because the inside is too dark and the outside too bright. (fig. p 58/59) In the old days

of photography you would simply lose something, whatever it was, and that loss was photography, it showed you just what photography was. And that was the beauty of photography.

Now, you can make a montage, as I did, where you don't lose anything. You can combine images and, in doing so, resolve the old problems, or at least seem to resolve them. So now you gain something. In the past, with photography as it was, your eye could tell you what escaped the photograph. If you were actually at the Barcelona pavilion on a sunny day, you could see the outside and the inside together at the same time because our eyes can do that. And so we learned that photographic film is more limited than our eyes are. We could see that the photograph could not really show us what our eyes saw. We recognised the limitations of photography but we experienced those limitations as what is beautiful in photography. But what our eyes saw was also beautiful. And now that you can capture that as well, it is also beautiful, maybe in fact as beautiful as the old photography. So therefore we have gained a little bit. We haven't lost anything, because we are not prevented from photographing in a traditional way. You can still choose to lose something if you want to.

Herzog

The way you describe the two methods of working — traditional photography and simulative, computer-aided photography — their potential for interaction offers prospects that could undoubtedly apply to architecture and city planning as well: being able to engage in one activity without losing or surrendering the other. In the case of our Chinese urban project in Jinhua, we're using simulative methods in order to bring together concrete and familiar urban spaces and building typologies from all over the world in a kind of genetic operation and then join them up to create a new whole. (fig. p 35) So you come up with an artificial solution. And, as in any other city, that solution is subject to transformation in the process of concrete, physical implementation and the way people use it in everyday reality. As I said before, it's going to be extremely interesting to see how this process of transformation develops and we want to give it as much leeway as possible.

Wall

Your model would lead to a plan and the plan would be executed imperfectly.

Herzog

Yes, inevitably: the plan will be imperfect, the execution of the plan will be imperfect, and life in the new city will be imperfect. But it will be an attempt to accomplish this imperfection on the highest level possible.

Wall

The utopian aggression against the actual, against the slow and the imperfect — I see that as a rhetoric, as one of the last formations of the avantgarde. Democracy involves imperfection. The fundamental aesthetic trait of democratic culture is the taste for imperfection. It has to do with accepting its presence and of knowing that everything you do won't be realised exactly as you want it to be, and that other people will also have something to say about it. That spirit of imperfection realises that the past was made from mistakes that we now find interesting, as interesting as 'getting it right' might have been, maybe more interesting.
For an artist making pictures, that beautiful imperfection is relatively easy to attain, much more so than for someone modelling a new city.

Herzog

When you compare existing examples of urban development, Switzerland is a relatively successful model of dealing with imperfection. Switzerland has pushed the imperfect and the fragmented to perfection.

Ursprung

But then you have to accept "art lite" or "architecture lite" as an inevitable part of democracy.

Wall

We have to accept it, but we don't have to accept it uncritically, because criticism is part of democracy, too. Acceptance includes acceptance of dissent. So if any of us dissent from "art lite", we are in some way accepting it or at least accepting the fact that we relate to it in the form of dissent. The mediocrity that surrounds us — we hope it is a surrounding mediocrity and we are not part of it! — is a phenomenon of the democratic spirit of our wasteful, prosperous society. It's not bad, as such, and, in fact, a lot of it is fun and amusing, fashionable and entertaining.

Ursprung

You are both in a situation of power in the sense that you've been able to create a lot, you've even created schools following in your wake. From your point of view this mediocrity might look amusing. On the other hand, many people admire your art and your buildings precisely because they see them as something that offers meaning, that provides protective spaces and that promises to change the surrounding atmosphere of powerlessness.

Wall

I don't think that art has anything new to offer. Powerlessness and fear aren't new. We can respond to them with only our discipline, our craft and our judgement. An architect probably doesn't have anything else, either, but probably has more ability to make an impact on others, maybe on society. As I said, the architect has to accept that many who experience his or her buildings will do so without any enjoyment or aesthetic appreciation, and they will still experience it legitimately. That's sort of tragic. Artists feel that anyone who doesn't enjoy their work does not really experience it. So we are insulated, we have this happy space of ours. But we cannot shape very much and so we do not have much direct effect on the affairs of the world. From within our space, our métier, we can contemplate and reflect on the difficulty, the burden, the obligation accepted by those who take on practical tasks. What we artists do is to observe what it is like to be burdened by practicality, acting as if we aren't.

Bechtler

Art is also a seismograph of a certain social reality and it's an instrument that can pose essential, critical questions of social and political significance. Art is like a delicate membrane that senses and responds to the vibrations of the times.

Herzog

Architects don't have much influence on the course of the world, but they build a stage for all those people about whom Jeff rightfully says that they use and perceive it with more or less enjoyment. As I have tried to explain, our burden is not "practicality" but rather the incredible leeway that seems to be growing all the time and yet, paradoxically, constricts because the coercion to be innovative threatens to nip all innovation and experimentation in the bud.

We can't change the situation. Our tools are limited. In China with the new resources, we can produce with the speed that is required of us but, on the other hand, medieval craftsmanship could almost be at the basis of this whole production: that's something quite new for me … And, in the final analysis, what counts won't be a city or its architecture, but how people live there and how close we have come to what we envisioned. … and whether anything can be a model for contemporary China in order to replace or be an alternative to destroying entire cities, which they are currently doing. That I think is something which is specific to architecture as opposed to art. However, for you as a photographer, it might be a new market to see how these people work.

Wall

I am sure photographers will be flocking there.

Biographies

Jacques Herzog, architect, born in Basel, Switzerland in 1950. He studied architecture at the Swiss Federal Institute of Technology Zurich with Aldo Rossi and Dolf Schnebli from 1970 to 1975 and began his partnership with Pierre de Meuron in 1978. He worked as an artist himself and had several one-man shows. He has been teaching in universities in the USA and Switzerland since 1989. Herzog & de Meuron have built numerous buildings and are currently constructing the National Stadium for the Olympic Games 2008 in Beijing, China. Many exhibitions and publications have been devoted to their work, and in 2001 they won the Pritzker Architecture Prize.

Jeff Wall is an artist working in photography. He has exhibited his pictures internationally for the past 25 years. Recent solo exhibits include Jeff Wall Tableaux at the Astrup Fearnley Museum for Modern Art, Jeff Wall Photographs at the Museum Moderner Kunst, Vienna, and Jeff Wall Landscapes at the Manchester City Art Gallery. Recent group exhibits include *The Last Picture Show: Artists Using Photography 1960 – 1982* at the Walker Art Center, Minneapolis, *Looking In — Looking Out: Positions in Contemporary Photography* at the Kunstmuseum in Basel and *Documenta 11* in Kassel. In 2003, Jeff Wall was awarded the Roswitha Haftmann Prize for the Visual Arts in Zurich, and in 2002 he received the Hasselblad Prize for Photography. Jeff Wall lives and works in Vancouver.

Philip Ursprung, art historian, born in Baltimore, USA in 1963. He studied in Geneva, Vienna and Berlin. He completed his post-doctoral thesis (Habilitation) at the ETH Zurich in 1999, and taught art history at universities in Switzerland and Germany. In 2001 he was appointed Swiss National Science Foundation Professor for Art History at the Institute for the History and Theory of Architecture at ETH Zürich. He is the curator of the exhibition *Herzog & de Meuron: Archaeology of the Mind* (Canadian Centre for Architecture, Montreal, 2002) and the editor of the exhibition catalogue *Herzog & de Meuron: Natural History* (Lars Müller, Baden, 2002). He is the author of *Grenzen der Kunst: Allan Kaprow und das Happening, Robert Smithson und die Land Art* (Silke Schreiber, Munich, 2003).

Cristina Bechtler, publisher, is founder and director of Ink Tree Editions, Küsnacht, Switzerland, an imprint for artist's books, editions and portfolios with contemporary artists. She conceived the book series *Art and Architecture in Discussion* in collaboration with the Kunsthaus Bregenz, Austria. Previous books in this series have been published with Frank O. Gehry/Kurt W. Forster; Rémy Zaugg/Herzog & de Meuron; and Mario Merz/Mario Botta. A roundtable discussion with Hans Ulrich Obrist, Beatrix Ruf, John Armleder, Didier Fiuza Faustino, Philip Ursprung, Jacques Herzog and Rem Koolhaas is in preparation.

Jeff Wall, Cristina Bechtler, Jacques Herzog, Philip Ursprung,
November 4th 2003, office of Herzog & de Meuron, Basel

List of Illustrations

p 8
Jeff Wall, Cristina Bechtler, Jacques Herzog, Philip Ursprung,
November 4th 2003, office of Herzog & de Meuron, Basel

p 12
Herzog & de Meuron, Lego House, contribution to the exhibition
"L'architecture est un jeu ... magnifique", Centre Georges Pompidou, Paris,
July 10th to August 26th 1985, film still

p 14
Jeff Wall, *Dominus Estates Vineyard, Yountville, CA, Winery building by
Herzog & de Meuron, Basel*,
1999, black and white photograph, 209 x 267 cm

p 16
Herzog & de Meuron, Dominus Winery, Yountville, Napa Valley, California,
USA, 1998

p 18
Herzog & de Meuron, Eberswalde Technical School Library, Eberswalde,
Germany, 1999

p 20
Jeff Wall, *No*, 1983, transparency in lightbox, 330 x 229 cm

p 22
Herzog & de Meuron, Signal Box, Auf dem Wolf, Basel, Switzerland, 1994

p 24
Herzog & de Meuron, REHAB Basel, Centre for Spinal Cord and Brain Injuries, Basel, Switzerland, 2002

p 26
Herzog & de Meuron, Forum 2004, Building and Plaza, Barcelona, Spain, 2004

p 28
Jeff Wall, *The Stumbling Block*, 1991, transparency in lightbox, 229 x 337 cm

p 32
Jeff Wall, *The Crooked Path*, 1991, transparency in lightbox, 119 x 149 cm

p 34
Herzog & de Meuron, Tree Village Campus, Beijing, China, masterplan 2003–2004

p 35
Herzog & de Meuron, Jidong New Development Area, Jinhua Commercial Cultural and Entertainment Center, Jinhua, China, project 2003–2004

p 36
Jeff Wall, *Outburst*, 1989, transparency in lightbox, 229 x 312 cm

p 37
Jeff Wall, *Mimic*, 1982, transparency in lightbox, 198 x 229 cm

p 38–39
Jeff Wall, *Restoration*, 1993, transparency in lightbox, 119 x 490 cm

p 40–41
Jeff Wall, *Dead Troops Talk (A Vision After an Ambush of a Red Army Patrol Near Moqor, Afghanistan, Winter 1986)*, 1992, transparency in lightbox, 229 x 417 cm

p44
Herzog & de Meuron, Ricola Storage Building, Laufen, Switzerland, 1987

p 50–51
Herzog & de Meuron, National Stadium, The Main Stadium for the 2008 Olympic Games, Beijing, China, completion projected for 2007

p 52
Jeff Wall, *The Destroyed Room*, 1978, transparency in lightbox, 159 x 234 cm

p 54
Herzog & de Meuron, Blue House, Oberwil, Switzerland, 1980

p 58–59
Jeff Wall, *Morning Cleaning, Mies van der Rohe Foundation, Barcelona*, 1999, transparency in lightbox, 206 x 370 cm

p 73
Jeff Wall, Cristina Bechtler, Jacques Herzog, Philip Ursprung, November 4th 2003, office of Herzog & de Meuron, Basel

Series Editor:
Cristina Bechtler, INK TREE, Seestrasse 21, CH-8700 Küsnacht
T. +41 1 910 71 76, F. +41 1 910 67 90, www.inktree.ch
in collaboration with:
Kunsthaus Bregenz, Karl-Titzian-Platz 1, A-6900 Bregenz
T. +43 5574 48 59 40, F. +43 5574 48 59 48, www.kunsthaus-bregenz.at

Translation: Catherine Schelbert, CH-Hertenstein

Editing: Philip Ursprung, Cristina Bechtler, Salome Schnetz

The printing of this publication was funded by Holcim Ltd., Zellweger Luwa Ltd., Burgauer Stiftung.

This work is subject to copyright.
All rights are reserved, whether the whole or part of the material is concerned, specifically those of
translation, reprinting, re-use of illustrations, broadcasting, reproduction by photocopying machines
or similar means, and storage in data banks.

The use of registered names, trademarks, etc. in this publication does not imply, even in the
absence of specific statement, that such names are exempt from the relevant protective laws and
regulations and therefore free for general use.

© 2004 for reproduced works by Jeff Wall
© 2004 Springer-Verlag/Wien and Authors
Printed in Austria
Springer-VerlagWienNewYork is a part of Springer Science+Business Media
springeronline.com

Photo Credits:
Jeff Wall (cover, p 14, 20, 28, 32, 36, 37, 38–39, 40–41, 52, 58–59), Herzog & de Meuron
(p 12, 34, 35, 50–51), Margherita Spiluttini (p 16, 18, 22, 24, 44, 54), Hisao Suzuki (p 26),
Peter Schnetz (p 8, 73)

Layout: David Marold / Springer Verlag KG, A-1201 Wien
Printing: Holzhausen Druck & Medien GmbH, A-1140 Wien

Printed on acid-free and chlorine-free bleached paper
SPIN: 10967799

Library of Congress Control Number: 2004105255

With numerous (partly coloured) Figures

ISSN 1613-5865
ISBN 3-211-20349-4 Springer-Verlag Wien New York

Previous publications:
- Discussion with Mario Botta and Mario Merz (out of print)
- Discussion with Herzog & de Meuron and Rémy Zaugg (out of print)
- Discussion with Frank Gehry and Kurt W. Forster (out of print)

In preparation:
- Discussion with Vito Acconci and Kenny Schachter
- Discussion with John Armleder, Didier Fiuza Faustino, Jacques Herzog, Rem Koolhaas, Beatrix Ruf, Philip Ursprung, moderated by Hans Ulrich Obrist